Hi! My nam
I'm happy †
Meet my wife, Marina, and
Marco and Misty, too.

We welcome you to Mystic,
our historic, seaside town.
There's a lot to see and do here,
and we'll show you around!

Every morning our whole family starts the day by taking wing. We go flying over Main Street to see what's happening.

We see visitors from everywhere having fun in town.

Then we stop and watch the bridge
as it goes up and down.

Our friend Denison is careful.
He closes every gate.
Until the ships pass safely through,
the traffic has to wait.

Mystic

Next, we're off to the Aquarium.
We might see you there!
Veterinarians treat the animals
with lots of love and care.

When we take our kids to this
world-famous destination,
we learn how to protect our planet
through ocean conservation.

At the docks in Stonington, fishing starts at dawn. Marco swoops by every day to see what's going on.

We shop at Olde Mistick Village,
and what we always do
is go and say "Hi!" to the ducks.
They can't wait to meet you!

9

The Charles W. Morgan is at the Seaport.
It's the oldest whaling ship in the world.
Marina thinks it's beautiful
with all its sails unfurled.

In the museum, we discover
the many different ways
boats and ships were built here
in the old seafaring days.

At the Submarine Base, we greet my sister, Skye, and brother, Davy. We're very proud of them because they're serving in the Navy.

Submarines stay underwater
for six months, sometimes more.
We cheer and welcome everyone
when they head back to shore.

13

Every summer, the Art Festival
is outdoors and it's free.
It's filled with paintings, arts and crafts,
and lots of pottery.

The Lighted Boat Parade is one of Misty's favorite nights when boats glide down the river covered in sparkling lights.

So, whenever you're in Mystic, look up in the sky.
You might see our family as we go flying by.
When you watch us soar above the sails, the sand, and sea
you'll know deep in your heart, right here...

is the best place to be.